A FIRST LOOK AT ANIMALS THAT EAT OTHER ANIMALS

By Millicent E. Selsam and Joyce Hunt

Illustrated by Harriett Springer

WALKER AND COMPANY ✺ NEW YORK

First published in the United States of America in 1989
by Walker Publishing Company, Inc.

Published simultaneously in Canada by Thomas Allen & Son
Canada, Limited, Markham, Ontario

Library of Congress Cataloging-in-Publication Data

Selsam, Millicent Ellis, 1912-
A first look at animals that eat other animals / by Millicent
Ellis Selsam and Joyce Hunt ; illustrated by Harriett Springer.
p.   cm.—(A First look at series)
Includes index.
Summary: A beginner's guide to distinguishing between different
carnivorous animals.
ISBN 0-8027-6895-4.—ISBN 0-8027-6896-2 (lib. bdg.)
1. Carnivora—Juvenile literature. [1. Carnivores.] I. Hunt,
Joyce. II. Springer, Harriett, ill. III. Title. IV. Series:
Selsam, Millicent Ellis, 1912-   First look at series.
QL737.C2S395   1989
599.74—dc20   89-33182
CIP
AC

Printed in the United States of America

10   8   6   4   2   1   3   5   7   9

A FIRST LOOK AT SERIES

LEAVES
FISH
MAMMALS
BIRDS
INSECTS
FROGS AND TOADS
SNAKES, LIZARDS, AND OTHER REPTILES
ANIMALS WITH BACKBONES
ANIMALS WITHOUT BACKBONES
FLOWERS
THE WORLD OF PLANTS
MONKEYS AND APES
SHARKS
WHALES
CATS
DOGS
HORSES
SEASHELLS
DINOSAURS
SPIDERS
ROCKS
BIRD NESTS
KANGAROOS, KOALAS, AND OTHER ANIMALS WITH
    POUCHES
OWLS, EAGLES, AND OTHER HUNTERS OF THE SKY
POISONOUS SNAKES
CATERPILLARS
SEALS, SEA LIONS, AND WALRUSES
ANIMALS WITH HORNS
ANIMALS THAT EAT OTHER ANIMALS

Each of the nature books in this series is planned to develop the child's powers of observation—to train him or her to notice distinguishing characteristics. A leaf is a leaf. A bird is a bird. An insect is an insect. That is true. But what makes an oak leaf different from a maple leaf? Why is a hawk different from an eagle, or a beetle different from a bug?

Classification is a painstaking science. These books give a child the essence of the search for differences that is the basis for scientific classification.

The authors wish to thank Mr. Kenneth Chambers of the American Museum of Natural History for reading the text of this book.

All animals eat to live.
How do they get their food?

Animals can't go to a refrigerator or to a supermarket.

They must find their food where they live.

Some, like lions and zebras, live
on the grassy plains of Africa.

Zebras eat grass.

Look at the zebra's teeth.
They are flat and can grind the tough grass.

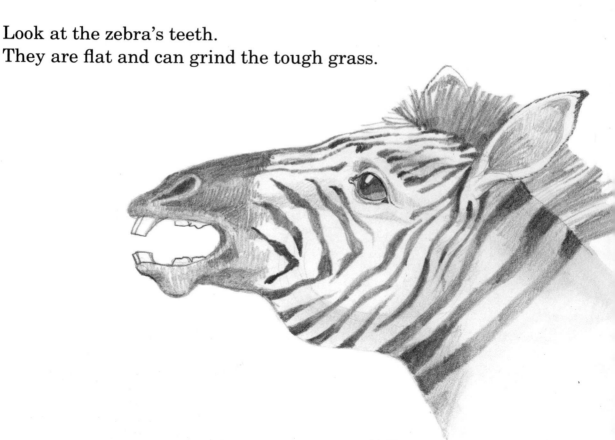

Lions eat zebras.

Look at the lion's teeth.
They are sharp and can cut the flesh of the zebra.

Lions and other animals with sharp, pointed teeth
and powerful jaws are called carnivores (<u>KAR</u>-ni-vawrs).
How do you tell carnivores apart?

Some are big.

Some are small.

Some are striped.

Some are spotted.

Some have long tails.

Some have short tails.

CATS

Lions are members of the cat family.
So are *leopards, pumas,* and *housecats.*

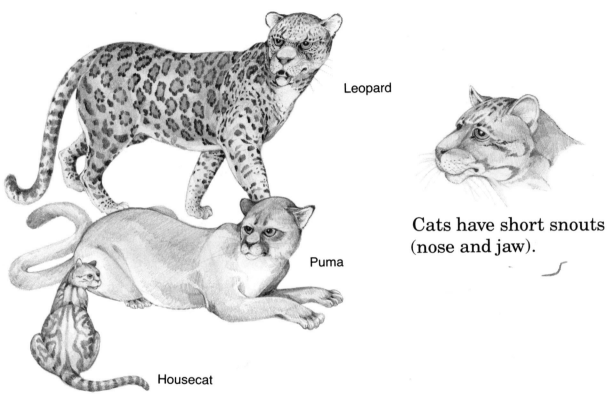

Leopard

Puma

Housecat

Cats have short snouts
(nose and jaw).

And most cats have claws that can
be pulled back into their paws.

Many cats are spotted or striped.
Match the cat to its pattern.
Cheetahs have spots.
Tigers have stripes.

Most cats have short legs
and hunt by stalking or creeping up on
their prey (the animals they eat).

DOGS
Most dogs have long legs and hunt
by running after their prey.

Wolves, coyotes (KI-oh-tees), and *foxes*
are all members of the dog family.

The largest member of the dog family is the *gray wolf*.

The smallest is the *fennec fox*.

The coyote is a small wolf.
Look at its long snout.
Most dogs have long snouts.

BEARS

Bears have large heavy bodies,
short powerful legs, and short tails.
They are the largest carnivores.

The huge polar bear can weigh as much as ten wolves.

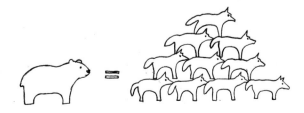

The grizzly bear is another heavyweight.

Smaller bears can be told apart by their markings.

Which bear has a white V on its chest?

Which one has a white U around its neck?

Which one has white circles around its eyes?
(It seems to be wearing sunglasses.)

Asiatic Black Bear

Sun Bear

Spectacled Bear

RACCOONS

Raccoons look like small bears with long, ringed tails.
They have black masks.

These two animals are in the raccoon family.

Find the *coati* (KOH-ah-tee).
It has a long nose.

Find the *kinkajou*.
It has a long tongue!

Some scientists think the *giant panda*
belongs in the bear family.
Other scientists think it belongs in the raccoon family.

A panda is a carnivore because of its teeth,
but it does not eat meat.

WEASELS

Most weasels have long thin bodies and very short legs.

The best known member of the
weasel family is the skunk.

It can shoot a powerful, smelly spray
that makes an enemy run the other way.

The smallest member of the weasel family
and the smallest carnivore in the world is called
a *least weasel*.
It can weigh as little as three walnuts.

Minks are large weasels.

Otters are among the largest members of the weasel family.

River otters swim and get their food in streams.

Sea otters are larger. They live in the oceans.

HYENA
The *hyena* (HI-ee-nah) has front legs that are longer than its back legs.

Its call sounds like a laugh.
Some people call it the "laughing" hyena.

Are people carnivores?
We eat meat.
But we do not have the sharp, pointed teeth
and the powerful jaws of a carnivore.

Many of the animals in this book are trapped and killed
because of their beautiful fur. What do you think about that?

To tell carnivores apart:
Look for short snouts and short legs.

Look for long snouts and long legs.

Look for big heavy bodies
and short tails.

Look for smaller bodies and
long, ringed tails.

Look for long, thin bodies and very short legs.

Look for long front legs and short back legs.

Look at the size.

Look at the markings.

CARNIVORES IN THIS BOOK